CAREERS IN
METEOROLOGY
AND
ATMOSPHERIC SCIENCE

METEOROLOGY IS THE STUDY OF THE atmospheric conditions that cause weather on earth. Most of these conditions occur in the troposphere, the layer of the atmosphere closest to the earth. Meteorologists can predict future weather conditions by studying patterns in temperature, air pressure, and water vapor. The media, private sector companies, and government agencies use these predictions to manage air and ocean traffic, predict crop yield, budget water, and in many other important ways.

Weather is everywhere and so are meteorologists. These professionals can be found all over the world doing all sorts of interesting things. Some study the ozone layer and look for ways to prevent air pollution or global climate change. Some monitor rainfall and issue flash flood warnings, or fly in specialized aircraft to investigate hurricanes. Most work for government agencies, such as the National Weather Service, providing vital information to the public as well as the aviation, marine, and fire control communities. Beyond the government, the fastest growing area for meteorologists is private forecasting. Private forecasters serve clients with very specific needs for highly specialized forecasts. For example, they might work for commodities traders who want to know how the weather will affect future crop production and prices. They might keep utility companies informed about impending hot or cold weather that will put heavy demands on generating plants and transmission systems.

Weather forecasting is at the heart of meteorology. The weather forecast that you get in your hometown is the end product of a worldwide effort by thousands of meteorologists in many nations. All those meteorologists use tools such as Doppler radar, satellites, and instruments that take precise atmospheric measurements to follow and analyze the huge systems that will eventually bring us our local weather.

To be eligible for most entry-level jobs as a meteorologist, you will need to have at least a bachelor's degree in meteorology or a related field. Along with the degree, you will need some experience pertaining to meteorology and related disciplines, such as thermodynamics, climatology, and even statistics and chemistry. Most people get that experience through student training programs and internships. Some are fortunate enough to find employers that offer on-the-job training either in-house or in the field.

Aspiring meteorologists can expect favorable job prospects, especially in private industry. The federal government will still be the largest single employer, with particular emphasis on research related to global climate change. Considering the economic impact of weather – an estimated $3 trillion a year – it is not surprising that the fastest job growth will be in private industry. The opportunities for weather broadcasters are limited and highly competitive.

Meteorology is a good choice for anyone with a passion for weather events, a head for math and science, and a desire to do work that benefits others. It is routinely ranked among the best jobs in America because it offers job security, little stress, plenty of employment options, and excellent compensation. There are numerous rewards for anyone with the sound knowledge of meteorology and the ability to use it in atmospheric research or applied meteorology.

WHAT YOU CAN DO NOW

METEOROLOGISTS NEED TO STUDY math and science. In high school, recommended science courses include physics, chemistry, earth science, and computer science. Earth science is of particular importance because that is where you will get a valuable introduction to the atmospheric environment. Proficiency in math is very important. Computer literacy is also essential. A good command of written and spoken English, and some foreign language experience will help you communicate scientific knowledge effectively.

Ask your school counselor to help you arrange job shadowing. This is a common activity for high school students checking out the careers that interest them. It is extremely valuable in helping make that important decision about whether a career in meteorology is really what you want to pursue. You will be able to observe a meteorologist at work and see what this career is all about.

The National Weather Service has locations all across the country. Some are national support centers, some are regional support centers, and many more are local forecast offices. Go to the NWS website to find the office nearest you. It is highly recommended that you visit an NWS office. You will be welcome and can learn about the jobs there. Ask to talk about this career path with the Meteorologist-in-Charge (MIC) or a forecaster. You can ask about meteorology work in general, or more specifically about career opportunities with the NWS. This way, you can get a real feel for the field.

The NWS also has student volunteer and employment opportunities available to help you gain experience and exposure in the NWS. Call your local office for details.

HISTORY OF THE CAREER

HUMANS HAVE HAD A VESTED interest in weather prediction for thousands of years, since the shift from nomadic to agricultural societies. When crop yield depended on predictable precipitation patterns, understanding the weather became a matter of survival.

The term meteorology comes from the Greek word *meteoron* – meaning something that happens high in the sky. The scientists of ancient Greece studied the clouds, winds, and rain and theorized how they were all connected. The weather was important in their relatively simple society because it affected how much food could be produced on farms, and whether sailors could safely traverse the oceans. Aristotle, considered the father of meteorology, wrote the first book on the subject. *Meteorologica*, a treatise that described weather patterns and other observations of atmospheric conditions, was written around 340 BC.

All societies throughout history tried to understand and predict weather patterns. Chinese weather prediction lore can be traced back as far as 300 BC. Indian astronomers developed their own methods of weather prediction around the same time. The ability to forecast rain helped the Persians produce more agricultural products.

Most ancient weather forecasting methods were based on pattern recognition, so named because it relied on the observed patterns of events such as wind movements or lunar phases. For example, the old sailor's adage "red skies at night, sailor's delight" arose from the observation that a red sunset was typically followed by a day of fair weather. These kinds of observations accumulated over the generations to produce common weather lore. Not all of the predictions proved reliable, but many did withstand the test of time.

Eventually, scientists realized that speculation and observation needed to be backed up by measurable facts that could be recorded and analyzed. For a very long time, the only features of the weather that could be measured were wind direction and rainfall. It was not until the 17th century that the basic tools of weather forecasting, the thermometer and the barometer, were invented. Over the next 200 years, more devices were developed to measure wind speed, humidity, and other features of the atmosphere. By the 1800s, scientists were using a wide array of instruments to record the long-term trends that are now known as climate.

The age of modern weather forecasting began with the invention of the electric telegraph in 1835. Until this time, there was no practical way to disseminate weather predictions on a daily basis. By the late 1840s, meteorology was flourishing, because information could be conveyed over great distances via the telegraph nearly instantaneously. This rapid exchange of information made it possible for forecasters to predict the weather in one location based on the knowledge of conditions further "upwind," generally meaning to the west. Future weather over much of the United States and Europe could be predicted by watching storms develop and assuming that they would move eastward.

The two men often credited with transforming forecasting into a science were Francis Beaufort and his protégé Robert Fitzroy. The Beaufort scale measured wind force and the Fitzroy barometer measured air pressure. Together, these instruments allowed the men to take numerous measurements that were used to forecast short-term changes in the weather with a high degree of accuracy.

In the early 1900s, a group of Norwegian meteorologists developed the foundation of modern weather forecasting. They started to apply basic laws of physics to their study of the behavior of the atmosphere. They observed movements of huge cold and warm air masses and the "fronts" where they meet, and concluded that local weather conditions were actually part of much larger weather systems that could cover hundreds or even thousands of miles.

The science of meteorology made great strides during the 20th century. In 1922, Lewis Fry Richardson discovered that mathematical calculations could be applied to weather prediction. However, the size and complexity of these calculations would require a computer – something that did not exist yet. It was not until 1950 that the first computerized weather forecast was performed. Within a few short years, the development of programmable computers had made numerical weather prediction practical.

The first public radio forecasts were made in 1925 by Edward B. Rideout, an employee of the US Weather Bureau. Television forecasts did not come along until 1947. The 24-hour cable network, The Weather Channel, began broadcasting in 1982.

Some of the greatest advances in meteorology arose out of necessity during World War II. Large-scale military land, sea, and air campaigns were highly dependent on weather over vast regions of the world. University meteorology departments were forced to expand rapidly as the military services sent cadets to be trained as weather officers. The military also supported scientific research that led to the development of radar and high performance aircraft. These wartime technological developments could probe the violent cores of thunderstorms, observe hurricanes and track fronts, and produce basic information needed for accurate forecasting.

The age of modern meteorology came in 1960, with the launch of the first successful weather satellite. The TIROS-1 collected and sent nearly 23,000 images of atmospheric conditions from space. Today, earth-observing satellites provide current information about land and water temperatures, cloud systems, and wind patterns. The information is processed by supercomputers to analyze and predict the behavior of the atmosphere on every scale, from the formation of raindrops to the circulation of the atmosphere over the entire earth. With continuing advances in satellite technology and communications, meteorologists may soon be able to predict natural disasters such as tsunamis with enough advance notice to minimize damage and prevent human loss.

Several government agencies use meteorological information to monitor conditions both on earth and in space. The National Oceanic and Atmospheric Administration (NOAA) founded in 1970, is responsible for anticipating and responding to climate and its impact on the planet. The National Weather Service is a branch of NOAA that provides services such as forecasts, severe weather advisories, warnings and watches. The National Aeronautics and Space Administration (NASA) uses information from the Earth Observing System to monitor global issues such as ozone depletion.

Our complex society and the environment are seriously affected by events and changes in the atmosphere. Weather forecasts can do far more than help you decide whether to carry an umbrella to work. Meteorology is an ancient science that has matured into a technologically advanced field of research. Meteorologists seek to address many complicated issues and answer many difficult questions about the behavior of the atmosphere and its effects on people and the planet.

WHERE YOU WILL WORK

METEOROLOGISTS WORK IN GOVERNMENT agencies, private consulting and research services, industrial enterprises, utilities, radio and television stations, and in education. Since weather is everywhere, meteorologists can be found all over the world doing many different jobs.

Most people are first introduced to meteorology through local television stations – and it may seem like this is the only place meteorologists work. It is true that many stations employ professional meteorologists rather than reporters to present weather forecasts to their viewers. Actually, only about a thousand professional meteorologists work in the media. There are many other organizations that employ meteorologists, including engineering and environmental firms, private consulting firms, research organizations, educational institutions, and over a dozen federal government agencies.

Government Jobs

Here in the United States, by far the largest employer of meteorologists is the federal government. Most work for the National Oceanic and Atmospheric Administration (NOAA), which includes the National Weather Service. The National

Weather Service alone employs more than 5000 professionals. The majority of them work as weather forecasters in weather stations throughout the country – at airports, in or near cities, and in isolated and remote areas. In smaller stations, they often work alone, and in larger ones, they work as part of a team.

Some government-employed meteorologists are on active duty with the military services, primarily the Air Force and the Navy. Others are civilian employees of the Department of Defense, the National Aeronautics and Space Administration (NASA), the Department of Energy, and the Department of Agriculture.

Private Industry

One of the fastest growing areas for meteorologists is industry. Private sector meteorologists provide a variety of services to industries and other organizations. Some are consulting meteorologists with their own companies and others work for corporations. In recent years, a rapidly growing specialty in meteorology has been in the area of information services, where companies develop software and other computerized systems to provide highly specialized weather data and reports.

The government also contracts private firms. There are many small and large private organizations doing research at the direction of federal agencies. This is such a huge and growing area that there are professional directories where these companies advertise their atmospheric science capabilities and specialties.

The Air Force also contracts with commercial companies to provide additional weather forecasting and observing at certain Air Force bases.

Universities

More than 100 universities and colleges in the United States and Canada employ atmospheric scientists. In addition to holding a faculty or teaching position, university and college professors often perform research, typically supported by government or foundation grants. Research programs can be found on all kinds of campuses, from large state universities to small colleges to specialized institutions.

Work Environment

Most weather stations are located at airports or in large cities, although some are located in isolated areas. Weather station employees may watch actual weather conditions from the ground or from an aircraft. Atmospheric scientists involved in research often work in offices and laboratories, but fieldwork and travel are common. They may travel frequently to collect data in the field and to observe weather events, such as tornadoes, up close. Meteorologists working in private industry often travel to meet with clients or to gather information in the field. For example, forensic meteorologists may need to collect information from the scene of an accident as part of their investigation.

Broadcast meteorologists give their reports to listeners and viewers from television and radio studios. They may also broadcast from outdoor locations to tell audiences about current weather conditions.

Work Schedules

Most atmospheric scientists work full time on a standard 40-hour schedule. Weather conditions can change at a moment's notice, making it necessary to continuously monitor conditions.

Jobs at most weather stations require rotating shifts to cover all 24 hours in a day, including weekends and holidays. During severe weather, such as hurricanes, hours may be extended beyond the standard assigned shift.

THE WORK YOU WILL DO

YOU CAN FIND A METEOROLOGIST ANY day of the week, in your living room, sharing the predicted week's weather forecast on the evening news. Weather is everywhere and it can affect us in numerous ways. Meteorologists can be found all over the world doing many different jobs, such as:

Operational forecasters analyze weather conditions and issue severe weather warnings to the public.

Military meteorologists provide valuable long-range weather forecasts for missions around the world.

Private sector forecasters tell road crews how many snowplows to have ready for an on-coming blizzard.

Aviation meteorologists tell pilots what weather conditions to expect at the time of take-off, landing, and in flight.

Meteorologists alert city managers to prepare for flooding or tornadoes that are heading their way.

Meteorologists save crops by warning farmers to turn on sprinklers or smudge pots when frost is coming.

Consulting meteorologists tell power companies when a heat wave is coming so they have enough power ready to run air conditioners.

Basically, meteorologists study and predict the weather and climate, and its relationship to the environment, the economy, and people's lives. Many collaborate with scientists and professionals in other fields to help solve problems in areas such as commerce, energy, transportation, agriculture, and the environment. For example, some might team up with engineers to find the best locations for new wind farms to generate electricity. Others might work closely with hydrologists to monitor the impact of climate change, and devise ways to manage changing water resources.

Meteorologists use a variety of different tools and highly developed instruments to do their jobs. For example, weather balloons, radar systems, satellites, and sensors are used to monitor and collect weather data. Sophisticated computer models are used to analyze and understand air pollution, drought, loss of the ozone layer, and other problems.

The three most important tools of modern meteorology are radar, satellites, and computers. The best tool available for detecting tornadoes and other dangerous kinds of severe weather is Doppler radar. It also is the key element of the new wind-shear detection and warning system that is being used at major US airports.

Satellites use advanced remote-sensing techniques to measure temperature, winds, and other qualities of the atmosphere at many levels. They are among the most valuable tools of meteorology because they can cover the entire surface of the earth, including vast ocean areas where no weather stations exist, to monitor changes in global climate.

The one tool that revolutionized meteorology the most is the computer. With high-speed computers, meteorologists can simulate days, weeks, and years of atmospheric behavior in minutes.

For hundreds of years, observations of the atmosphere have been made on the ground at stations equipped with instruments to measure temperature, barometric pressure, humidity, and wind speed and direction. Today, there are special aircraft designed to fly in extreme weather, to analyze weather conditions up-close. These high-performance airplanes, equipped with measuring and sampling instruments, provide extremely accurate data. The "hurricane hunter" is designed to fly into the heart of the biggest and most intense weather systems. High-performance jets can fly into plumes of smoke and ash over erupting volcanoes to sample particles that are ejected into the atmosphere, where they can affect weather and climate.

Forecasters

Meteorologists work in a range of fields and typically specialize in one type of work. The largest group of specialists is forecasters. Also known as operational meteorologists, these professionals use computer and mathematical models to analyze current and expected weather conditions. They predict short and long range weather changes that can extend from a few minutes to more than a week, based on data received from satellites and weather stations around the world. General forecasters provide weather summaries for limited geographic areas. There are also specialized forecasters who develop forecasts for use in agriculture, aviation, forestry, and marine operations.

Forecasting has always been the principle job of meteorology. It involves many people in many countries, because weather systems are hundreds of miles in size and move across huge regions of the earth's surface as they grow and change. The weather forecast seen on your local television station is the end product of a worldwide effort by thousands of meteorologists in the national weather services of many nations.

Media Weathercasting

Broadcast meteorologists are the people who report and predict the weather for television, radio, newspapers, and the internet. They are responsible for gathering data, creating a forecast, and then using graphics software to display maps and charts that explain their forecasts to viewers. Media weathercasting is by far the highest profile of all careers in meteorology. These professionals often represent a strong link to the community, and they are frequently called upon to bring weather into school classrooms or city planning meetings. Sometimes they are asked to act as environmental reporters, generating stories on a variety of earth topics.

Obviously, strong communications skills are essential for this job, and a charismatic personality will help. A strong theoretical background in meteorology is a necessity along with forecast experience and computer competence.

Research

In general, the goal of atmospheric research is to improve basic understandings of climate and complex weather phenomena such as hurricanes, tornadoes, severe thunderstorms, snowstorms, and the dangers that accompany them. With better understanding, forecasters may improve their forecasts and save lives and property.

Most research meteorologists have a particular issue they are studying. Some are focused on environmental problems, such as air pollution, while others are studying microbursts so they can develop more accurate wind-shear detection and warning systems that will make air travel safer.

Research meteorologists often work with scientists in other fields, such as math, chemistry, physics, and hydrology. For example, meteorologists and oceanographers are working together to study many important ocean-atmosphere interactions such as El Niño.

Climatology

This type of meteorologist studies historical weather patterns and data to help predict future climate trends. They may collect and analyze past records of temperatures or rainfall, ranging over months, years, or even centuries. The data is used to interpret and forecast shifts in climate, usually for specific regions, years or decades into the future. Their findings are often used by economists and urban planners to design buildings, plan heating and cooling systems, and aid in efficient land use and agricultural production.

The largest area of study for climatologists today is global climate change. Scientists in this growing field, also known as earth systems science, collect and analyze past and present data on a worldwide scale. They study interactions among the atmosphere and the oceans, the polar ice caps, and the earth's plants and animals, to understand global temperature trends and changes.

Forensics

Whenever weather conditions have an impact on legal cases, forensic meteorologists may be called upon to provide expert knowledge. They are the CSIs of weather, investigating traffic accidents, suspicious fires, insurance fraud, and environmental regulatory violations. Acting as a background consultant, the forensic meteorologist will retrieve and analyze archived weather record information and reconstruct the weather conditions for the location and time of the event in question. They may be called to testify as expert witnesses in court. Consultation can be as simple as answering a quick question such as, "Was there lightning in the area when the house caught fire?" Or it can be as involved as preparing a comprehensive and site-specific analysis. ("Was the crash caused by poor visibility from natural fog or pollutants from the nearby industrial plants?")

Consulting

Companies large and small now turn to meteorologists to improve their bottom lines. A single pizza delivery store might bring in $500 more on a bitterly cold night – if they know in advance to schedule more help. On a larger scale, a city that knows the temperature will rise 10 degrees on a summer day can alert a local electrical utility to purchase extra power before the demand soars, and the prices of power go up with it. The correct forecast of a mere few degrees difference in temperature can save an electric company millions of dollars. A harsh winter can cost the entire US economy billions of dollars in direct economic losses due to lost retail sales, increased energy consumption, transportation problems, crop losses, and lower industrial production. Consulting meteorologists, also known as industrial forecasters, provide advanced warnings that can help reduce some of the losses.

Education

These experts use their knowledge of the atmosphere and meteorology to become educators. Atmospheric science education at the college and university level has grown tremendously in recent years. In addition to classroom teaching, many university atmospheric scientists conduct vital research on global climate change. They also direct research that graduate students are performing to earn their degrees.

STORIES OF REAL LIFE METEOROLOGISTS

I Am a Storm Chaser

"I was born and raised in the state of Arkansas, where I experienced almost every form of extreme weather except hurricanes. As a kid, I was always fascinated with floods and lightning. As I got older, it was tornadoes and derecho events (intense, long-lived windstorms). My first significant severe weather experience was a destructive hailstorm that struck my hometown. The very same year I experienced my first tornado, an F3 that caused more than a million dollars in damage to the town. That was the storm that really got me thinking about a career in weather.

While I was earning my Bachelor of Science degree in meteorology, I took some storm spotter training classes. Spotters keep a visual on a storm in their local area, although they don't chase. It requires the same basic knowledge, and the goal is the same – to call in observations to the National Weather Service during a storm. Those warnings from the field can help save lives. By the time I graduated from college, I had become really serious about chasing.

I'm primarily a Dixie Alley storm chaser, although I cover a lot of territory and sometimes go all the way out to Texas, Oklahoma, or Kansas. I am employed by a major TV cable news network, working as part of a team that provides live coverage to give viewers a very realistic sense of what is happening and where. I work with a camera crew, but mine is a real meteorology job, more than a reporter.

A professional chaser can cover 4,000 miles in a week! So far, I have chased over 100 events. It is hard to say which was the most memorable – there have been so many – but I would have to say the EF5 tornado in El Reno ranks near the top. There were so many things happening at once, it was overwhelming. The inflow winds were around 115 mph, and I saw numerous power flashes about half a mile in front of me. I kept adjusting my heading into the wind just trying to keep from being rolled. Then suddenly this monster tornado blows up about 300 yards away. It was a scary storm, but the people running away were even scarier. Cars kept coming right at me, trying to get away as fast as they could. It made it hard to maneuver so our crew could do our jobs safely. Remember, storm chasing isn't about having a thrilling joy ride. We have to document events and relay warnings to the National Weather Service to help save lives.

Storm chasing is a cool job, but you shouldn't consider it at all unless you are serious and learn a lot first because it can be very dangerous. It's about more than tornadoes, which can usually be avoided if you know what you are doing. All storms can be dangerous. Wet roads can cause a car to skid, floods can sweep cars right off the road, big hail can break windshields and cause serious injury, and lightning can be fatal.

I chase because I absolutely love to experience the sheer power and beauty of nature. Part of the appeal is that every storm is different, so each storm is a puzzle to be worked out. The thing I like most is helping people. So many times I have witnessed the devastation caused by severe weather. The work I do makes it possible for people to know when weather is potentially life-threatening so they can take the necessary steps to protect themselves and their families."

I Am a Fire Weather Forecaster

"My career in meteorology began with a summer internship at the National Weather Service office in San Francisco. When I completed my bachelor's degree in meteorology, the NWS offered me a full-time assignment to the fire weather office in Northern California. The station is a hub of fire weather activity because it provides meteorological support for fire management over the state and federal wild land of Northern California. I quickly learned how important good weather support is to safe and effective wild land fire management. I continued taking advanced training courses in fire weather, fire behavior, and fire danger rating.

My job duties require me to shift my meteorological focus between various forecast disciplines depending upon the time of year. During non-fire seasons, I spend most of my time working on spot forecasts. These are detailed predictions of microclimate weather conditions that exist in a specific forecast district. They are vital for helping fire managers safely meet the goals of prescribed fire management or wildfire suppression. During this time, I also investigate problem wildfire outbreaks through applied research.

It is during fire season that all of my knowledge and skills are needed most. The toughest part of the job is weather forecasting in complex terrain with minimal data support. I need to be able to recognize the components of severe wildfire burning conditions as well as weather patterns. I produce general fire weather forecasts for fire planning and issue Red Flag Warnings when necessary. A Red Flag Warning is a special bulletin that is issued when weather conditions (such as Santa Ana winds or dry thunderstorms) reach the dangerous potential for extreme wildfire starts that may quickly get out of control. It is a "must have" bit of data for field foresters, who must take specific action such as increasing the number of fire crews on duty, or closing public forests to industry and recreation.

Fire weather forecasting has provided me with a rewarding career in operational meteorology. The combination of working in the field and studying computer modeling has resulted in the steady improvement of my skills as a meteorologist. In addition, I have enjoyed many opportunities for travel. I have participated in fascinating fire weather seminars in China, worked as a fire weather forecaster in Australia, and helped train student fire forecasters for fire agencies in several states."

PERSONAL QUALIFICATIONS

THE MOST SUCCESSFUL METEOROLOGISTS are passionate about their work. It takes genuine enthusiasm for all things meteorology to get through the educational requirements that are heavy on the math and sciences. It also helps to have a natural head for these subjects because meteorologists routinely use calculus and statistics, as well as complex equations and formulas to develop computer models used to forecast weather.

Communications skills are important in this field. Speaking skills are not just relevant for broadcast meteorologists who present their forecasts to the general public. All meteorologists must be able to explain their forecasts and research to team members, supervisors, and clients, as well as the public. Most meteorologists work as part of a group. In order to contribute to the group, which typically includes professionals of many different disciplines, interpersonal skills will be essential. Writing skills are also needed for preparing detailed reports of forecasts and research.

Being detail-oriented is a must, but being organized is even more important because meteorologists continually deal with large volumes of information.

Excellent problem-solving skills will really help in your success. Solid critical thinking skills are needed to analyze the results of computer models. Successful meteorologists are able to think conceptually in multiple dimensions, and apply theoretical concepts to determine the most likely outcome.

ATTRACTIVE FEATURES

FOR THE PAST 10 YEARS, METEOROLOGY has been ranked among the top best jobs in the US. Not surprising considering it is a field that offers good pay, a pleasant work environment, little stress, few physical demands, and good job security (layoffs are rare). The compensation alone is reason enough to attract new applicants. The median salary is $90,000 not counting generous overtime pay and benefits. There are many other reasons to consider this field seriously for your future.

One of the big advantages of being a meteorologist is there are so many employment options. You can be involved at many levels ranging from taking observations to working on high end theoretical problems on supercomputers. You can find employment within many federal government agencies, the military, state and local government, universities, broadcasting, utilities, private industry, engineering consulting firms, or even through self-employment as a consultant.

You can live just about anywhere because there are offices, labs, and weather stations nationwide, even worldwide.

You can become an expert in the specialized area of meteorology that you find most interesting – computers, severe weather, environmental impact, economics, etc.

However you choose to practice your profession, you can rest assured that you will not be bored. It is rarely monotonous work. There may be T-shirt temperatures one day, and then a few days later it might snow. Nothing is as changeable as the weather. The fun is in trying to figure out what will happen, as well as seeing why something did or did not happen as expected.

UNATTRACTIVE FEATURES

METEOROLOGY IS NOT (YET) AN exact science. There is still considerable room for error. Storm fronts can move more slowly or quickly than expected, or be more or less severe than predicted. People not only want to know what to expect from the weather, but also what to do about it. Should they board up their windows? Where should they park their cars? How much food should they buy? Most days, getting it wrong does not matter all that much. However, in the case of natural disasters, it can mean the loss of property or even life. That is when the pressure to be right can be intense.

Although the job outlook is good overall, those hoping for a career in broadcasting may be disappointed. Openings at television stations are limited, and the competition for those jobs is tough. Most of the action is in private industry, and the federal government will continue to be the biggest employer.

For most meteorologists, the physical demands are minimal and safety is not an issue, but for some, it is a risky business. Military meteorologists are sometimes needed to gather weather in advance of the main ground forces. For paratroopers, it is the meteorologists who will parachute to the ground first to collect weather information. Then they relay that information back so the rest of the paratroopers can come down safely. Those in research can also face danger. Some fly into hurricanes to collect information, and others work with large tanks of hydrogen gas that are used to fill most weather balloons.

EDUCATION AND TRAINING

THE NUMBER OF NEW POSITIONS IN meteorology is expected to continue growing over the next decade. The job outlook is good, especially for students with advanced degrees in the atmospheric sciences who also possess computer-related skills.

Low-level jobs, such as collecting data or performing basic forecasting, do exist for those with only a high school education or certificate earned through a vocational or technical school program.

The vast majority of meteorologists and atmospheric scientists have four year college bachelor's degrees, and many have graduate level master's and doctoral degrees. A master's degree in atmospheric science can greatly enhance employment opportunities, earnings and potential for advancement. A master's degree in business administration (MBA) may be useful for meteorologists interested in working in private industry as consultants who help firms make important business decisions based on their forecasts. Across the board, there is growing job competition due to increased interest in this fascinating field – it would be shortsighted to ignore the advantage a good education provides.

The nature of most work in today's atmospheric sciences is often multi-disciplinary, meaning training in a related subject beyond traditional meteorology is almost essential. The demand for meteorology services is coming from nearly every industry and government sector. It is not uncommon for meteorology students to sign up for extra courses, double majors, or even advanced degrees in areas such as physics, computer science, electrical engineering, physical chemistry, statistics, ecology, horticulture, and hydrology. A strong foundation in computer sciences and applications is essential.

Professional meteorologists typically work in specific areas. Therefore, students should also take courses in subjects relevant to the particular specialization that is of most interest. For example, a student looking to become a broadcast meteorologist for radio or television should develop excellent speaking skills through courses in communications, speech, journalism, and related fields.

In this age of global climate change, a growing number of aspiring atmospheric scientists want to specialize in research. In this case, a student should plan to obtain a master's degree at minimum, and preferably a PhD in atmospheric sciences or a related field. Interestingly, most graduate programs do not require a bachelor's degree in atmospheric science specifically. An undergraduate degree in mathematics, physics, or engineering provides excellent preparation for graduate study in atmospheric science. Depending on the type of research, graduate students typically take courses in related subjects, such as oceanography and geophysics.

A large number of meteorologists – almost 35 percent – are employed by the federal government, many working for the National Weather Service. Because it is the largest single employer of meteorologists, all students should be careful to choose a college that offers the courses the federal government requires of job applicants. Most entry-level jobs in the federal government require a bachelor's degree and at least 24 semester hours of meteorology or atmospheric science, with courses in atmospheric dynamics, physical meteorology, and remote sensing of the atmosphere. Other courses that could make a job candidate stand out would include aeronomy (the study of chemical and physical phenomena in the upper atmosphere), physical hydrology, and engineering. Classes in computer programming are especially valuable because many atmospheric scientists have to write and adapt the computer software programs that produce forecasts.

The American Meteorological Society (AMS) lists around 100 graduate and undergraduate atmospheric science programs to choose from. These 100 programs are in atmospheric, oceanic, hydrologic, and related sciences. However, many schools also offer atmospheric science courses through other departments, such as physics and geosciences.

Another good option is to go into the Air Force. In the Air Force, you can get meteorology training plus plenty of hands-on experience in weather observing and forecasting. The Air Force experience and benefits will allow you to continue your education in a meteorology college program while in the Air Force or once your Air Force career ends.

Professional Certification

Although there are no licensing or certification requirements for working in this field, there are professional certifications that may be useful in advancing your career. The American Meteorological Society (AMS) offers two of them.

The Certified Broadcast Meteorologist (CBM) designation is for those working in television and radio. The CBM program was established to raise the professional standards in broadcast meteorology and foster a broader range of scientific understanding, especially with respect to environmental issues. Candidates for the program must hold a degree in meteorology from an accredited college, pass a rigorous examination, and have their work reviewed to assess not only technical competence, but also communications skills needed to be an effective broadcast meteorologist.

For consulting meteorologists, the American Meteorological Society offers the Certified Consulting Meteorologist (CCM) certification. With this program, the AMS has established high standards of technical competence, character, and experience for those providing advice and consultation in meteorology. The CCM designation indicates that holders have been tested and found to meet or exceed those standards. The essential attribute of the CCM is a specialized knowledge of the field, combined with broad experience, an integrated concept of service, and a clear and unwavering adherence to the rules of professional conduct and service.

Meteorologists with more experience than education might want to try for the National Weather Association's (NWA) certification. Applicants do not have to have a meteorology degree to qualify for the NWA Seal of Approval. However, the NWA requires two years of full-time experience.

EARNINGS

METEOROLOGISTS AND ATMOSPHERIC scientists generally receive good pay and benefits. The median annual salary for atmospheric scientists is around $90,000. Only a few earn less than $60,000, and the highest paid earn more than $135,000.

Median annual wages in industries employing the largest numbers of atmospheric scientists are:

Federal government
$95,000

Research and development
$90,000

Colleges, universities, and professional schools
$80,000

Radio and television broadcasting
$80,000

Professional, scientific, and technical consulting
$60,000

Most meteorologists and atmospheric scientists enjoy guaranteed substantial yearly pay raises in the first few years of employment. There is also compensation for work performed on holidays, at night, or on Sundays.

All full-time meteorologists and atmospheric scientists receive benefits. Typical benefits include health and life insurance, paid vacations, sick leave, and retirement pay.

OPPORTUNITIES

THE NUMBER OF JOB OPPORTUNITIES in this field is projected to grow by more than 10 percent over the next decade. That is considered average growth. The largest single employer will continue to be the federal government, along with the National Oceanic and Atmospheric Administration (NOAA).

Atmospheric scientists will always be needed to analyze and monitor air pollution and global weather observations. New technology is creating additional areas where their skills can be applied. Computer models have greatly improved the accuracy of forecasts, which means atmospheric scientists can now focus their forecasts on very specific purposes. This is increasing the demand for atmospheric scientists to work in private industry, as businesses find more ways to utilize specialized weather information. For example, farmers used to be one of the few groups of people whose business and income depended on the whims of Mother Nature. Long before computers and daily weather reports from local TV stations, farmers relied on *The Old Farmer's Almanac* for astronomical and weather predictions. First published in 1793, the Almanac has remained extremely popular among farmers and others who want a general idea of what to expect weather-wise from day to day. It has been surprisingly accurate – but not so accurate that brokers in the Chicago Board of Trade would consider it a basis for investing in corn futures – they use the services of meteorologists for that.

It is estimated that weather in the United States affects around $3 trillion in private industry. That includes everything from Target and Walmart registering disappointing sales due to too much snow, to ski resorts that lose money by low attendance when there is not enough snow. Today, meteorologists are able to predict weather to such a high degree of accuracy that all kinds of companies use the data to offset weather risk, predict consumer behavior, and hedge against weather-related losses through insurance policies known as weather derivatives. As more companies become aware of the increased accuracy in weather forecasting, the demand for the people who provide those forecasts will grow. In fact, the best option for most aspiring meteorologists is in the private sector working for corporations.

One non-business related area that offers many opportunities is the growing green movement. The focus on the current environmental climate crisis of global warming has everyone's attention. In recent years, there have been tremendous losses from erratic weather patterns, fierce storms, hurricanes, floods, and drought. This area holds a great deal of potential in the coming years.

The US Department of Defense also employs atmospheric scientists, and members of the Armed Forces are involved in meteorological work and forecasting. This is especially true for both the Navy and Air Force. Understanding and predicting weather patterns are equally important on sea as in the air.

The TV news media and the internet are booming industries, and some of the best future opportunities may be working for a local station or national television news network, like The Weather Channel, or a website like weather.com. The travel industry is also another promising area for employment, as weather patterns affect all aspects of travel.

GETTING STARTED

STARTING OUT AS AN INTERN CAN BE invaluable. Since the National Weather Service is the single largest employer of meteorologists in the US, you should start there. The NWS Forecast Office has two programs for high school or college students. The first is the Student Temporary Employment Program (STEP). This program offers temporary employment at a forecast office. The positions may be full time or part time, and can range from summer jobs to positions that last as long as you are a student. After completing the initial appointment, you may be eligible for permanent employment after successfully completing your education and meeting work requirements. The real bonus is you may be able to convert to a permanent position without having to compete with other applicants.

The other NWS program is the Student Volunteer Service. As the name implies, these opportunities are unpaid, but they provide invaluable work experience related to your academic program. You can get information about these and other internship programs directly from the NOAA job website or from the American Meteorology Society (AMS).

Many college departments post job announcements on the bulletin boards near the department office. Professional organizations such as the National Weather Service and the National Center for Atmospheric Research routinely publish their open positions. Employment announcements can also be found on the websites of the University Corporation for Atmospheric Research, and the American Meteorology Society.

Many private sector jobs are arranged through personal contacts, so networking is vital. If you have the training, good references, and healthy work ethic, chances are the work will find you. Asking people already in the business can be a source of job leads. Here in the United States, opportunities are everywhere. However, there are a few hot spots to consider:

Boulder, Colorado, is home to NOAA. This government agency hires thousands of meteorologists and professionals in related disciplines. The Boulder facility houses the Climate Diagnostics Center, where meteorologists constantly monitor environmental conditions to help provide critical data to citizens, city planners, and members of emergency teams. NOAA also operates a dozen other environmental research laboratories. The more well-known labs include the Atlantic Oceanographic and Meteorological Laboratory, which includes the Hurricane Research Division in Miami and the National Severe Storms Laboratory in Norman, Oklahoma.

Boulder is also home to The National Center for Atmospheric Research (NCAR). This facility is heavily involved in global climate change studies, but its research is extensive, covering many atmospheric science disciplines. One of NCAR's main objectives is to help create the next generation of researchers needed for global climate studies.

Columbus, Ohio is home to the Byrd Polar Research Center. This is a valuable source of potential jobs because the Center, named after the famed polar explorer, Admiral Richard E. Byrd, is recognized worldwide as being a top institute for polar research. Their main focus is on how the polar region affects the overall climate of the planet.

Houston, Texas is an area frequently subjected to unusual weather patterns, making it one of the best places to find meteorologist jobs. Portland, Oregon and Lawrence, Kansas also provide a range of job opportunities for meteorologists.

NASA is a hub of research activity. Most NASA meteorologists work at the Goddard Space Flight Center in Greenbelt, Maryland; the Langley Research Center in Hampton, Virginia; and the Marshall Space Flight Center in Huntsville, Alabama. However, the best opportunities can be found at The Goddard Institute for Space Studies (GISS) in New York City. The GISS, which works in cooperation with Columbia University, has become a leader in global change studies.

ASSOCIATIONS

■ **National Weather Service**
http://www.weather.gov/organization

■ **National Oceanic and Atmospheric Administration (NOAA)**
http://www.noaa.gov

■ **American Geophysical Union**
http://sites.agu.org

■ **National Weather Association**
http://www.nwas.org

■ **American Meteorological Society**
http://www.ametsoc.org

■ **University Corporation for Atmospheric Research**
http://www2.ucar.edu

■ **The World Meteorological Organization (WMO)**
http://www.wmo.int/pages/index_en.html

PERIODICALS

■ **Weatherwise Magazine**
http://www.weatherwise.org

■ **International Journal of Meteorology**
http://www.ijmet.org

■ **National Weather Digest**
http://www.nwas.org/digest

WEBSITES

■ **Weather Prediction Education**
www.theweatherprediction.com

■ **University of Oklahoma, Norman Campus, College of Atmospheric and Geographic Sciences**
www.ou.edu/content/gradweb/aud/current/masters /norman.html

■ **Iowa State University**
http://www.las.iastate.edu/discover /academics/phys/meteorology.php

■ **Saint Cloud University**
http://bulletin.stcloudstate.edu/ugb/programs/ahs.asp

■ **Joint Institute for Marine and Atmospheric Research (JIMAR)**
http://www.soest.hawaii.edu/jimar

■ **Lyndon State College**
http://www.lyndonstate.edu/degree-programs/atmosphe ric-sciences-meteorology

■ **Rutgers**
http://meteorology.rutgers.edu

■ **Florida Institute of Technology**
http://www.fit.edu/programs/ugrad/bs_meteorology